A SHARECROPPER'S DAUGHTER

Tells Her Family Story

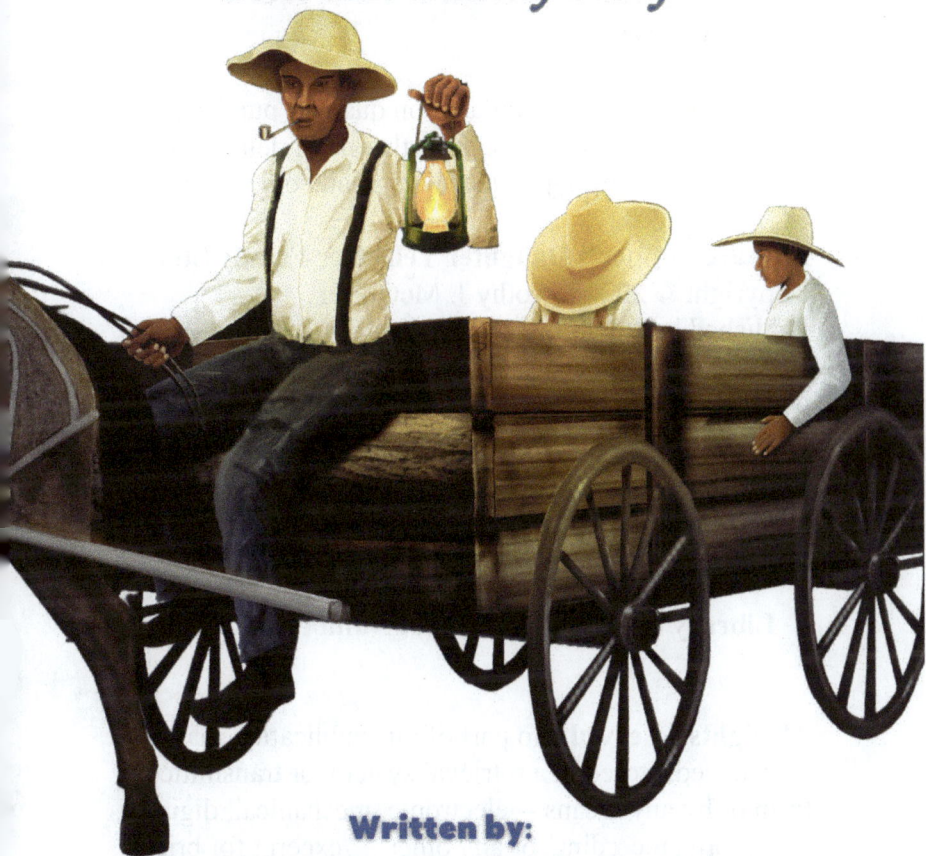

Written by:

Dorothy J. McQueen

INK TO LEGACY
PUBLISHING

Elgin, SC

Cover Design by *Gregory McQueen*
Edited by *Johnnerlyn Johnson*
Illustrations by *Mijah Simmons*
Pictures courtesy of the McQueen Family Album

Ink to Legacy Publishing
PO Box 555
Elgin, SC 29045

Special discounts are available on quantity purchases by
corporations, associations, and others. For details, contact
the author at dorothyjmcqueen7@gmail.com.

A Sharecropper's Daughter Tells Her Family Story
Copyright © 2024 Dorothy J. McQueen
Published by Ink to Legacy Publishing
Elgin, South Carolina 29045
www.ink2legacy.com

ISBN 979-8-9864280-6-2 (pbk)

ISBN 979-8-9864280-7-9 (ebook)

Library of Congress Control Number: 2025900323

Printed in the United States of America

Dedication

This book is dedicated to my parents, Alberta and George Jackson (deceased), my siblings Christine, Mary, Roberta, Georgia Ree, Paul, Frances, Margaret, William, and Columbus (all deceased), and Gloria, Ambrose, and Betty Jean (living).

Also, we cannot forget the thousands and thousands of sharecroppers who lived in the United States.

To the posterity... Today, your only stumbling block is yourself. Reach for the stars and achieve great things.

Table of Contents

Introduction

As the author of this book, it is my desire that the readers will conclude that the factual information presented is not intended to promote acrimonious feelings or dislike towards anyone. It is an honorable salute of dedication to my parents, Alberta and George Jackson, who reared thirteen children on a sharecropping farm for more than half a century.

Even through difficult times, they never abandoned their family. They made sure we had a roof over our heads and food on the table. They taught their children to be respectful and honest. Whatever jobs we were assigned, we were expected to do them well.

Each Sunday morning before we ate breakfast, we had to kneel and pray. After Papa led us in saying the Lord's prayer, he said another prayer. During that time, two persons' heads rested on the seat of chairs. About three persons were kneeling in front of an old trunk. Some were kneeling in front of the only bed in the room. Wherever we were, we had to kneel.

When I became a wife and the mother of four sons, I realized that my parents' faith in Jesus Christ was the reason they were able to survive.

In addition to this, profanity was never heard in our house. If there was such a beverage like beer and whisky, it was unknown in our house.

Also, Papa and Mama were respectful to the owners of the farm. We never heard them use unkind words about them.

Papa was an extraordinary farmer. He was only 5' 6" tall. His slim body would probably give someone the impression that he was a fragile man. However, he was the opposite. In fact, I have seen him several times lift a full sheet of cotton and throw it in the wagon.

Mama was about three or four inches taller than Papa. She was slim, and her hair was shoulder length. She liked for the girls to comb and braid her hair. She worked hard to help Papa on the farm. Also, she was an excellent cook.

During those years, there was no modern farming equipment. All the equipment used was drawn by a mule. When my four brothers became twelve years old, they helped Papa plow the fields and cut the firewood.

Not only did my father grow cotton, but he grew wheat for flour, corn for meal, peanuts, sugar cane (He made homemade molasses from sugar cane.), watermelons, cantaloupes, squash, okra, field peas, string beans, lima beans, collards, turnip greens, green onions, white potatoes, sweet potatoes, pumpkins, tomatoes, and beets.

I can still envision Papa on Saturday morning packing his wagon with fresh vegetables to be taken downtown to sell. After a few hours, we heard the empty wagon coming back to the house. When he got there, he gave Mama a bag with sugar in it and a few items she needed to help her prepare meals. I can imagine he bought himself some chewing tobacco too. Papa chewed tobacco until his health caused him to be confined to the bed.

By the time I was in high school, all the siblings had left the farm except four of us. Georgia Ree was an elementary school teacher. Frances was in her second year of college. Betty was getting ready to graduate from high school. Gloria, the youngest, was in the eighth grade.

During that time, we were making plans to move to the new house that our brothers built for Mama and Papa.

In 1961, we moved to the new house. The brick house had four bedrooms, a big kitchen, a dining room, a living room, one full bathroom, a front scoop, and a big side porch. A half bath and an enclosed back porch were added later.

Mama lived in the new house for two years. A massive heart attack ended her life. She was sixty-three years old. Christine, the eldest in the family, stated that Mama didn't want any of her children to die before she did. Her wish came true. She was the first in our immediate family to pass.

The date that Mama and Papa married is unknown. However, our eldest sister, Christine, who was the historian of the family, said that Mama was seventeen, and Papa was twenty-five when they were married. At the time of Mama's death, they would have been married for fifty-six years.

Papa lived in the new house for twenty-six years. During those years, he passed the written and driver's tests and got his license. He drove his car until his health failed him.

For approximately half of the time, he lived in the new house, Papa worked with his son, Columbus, who was a mason contractor.

On August 12, 1988, Papa passed at his residence after an extended period of illness. He was 93 years old.

Pictured above is Papa George Washington Jackson.

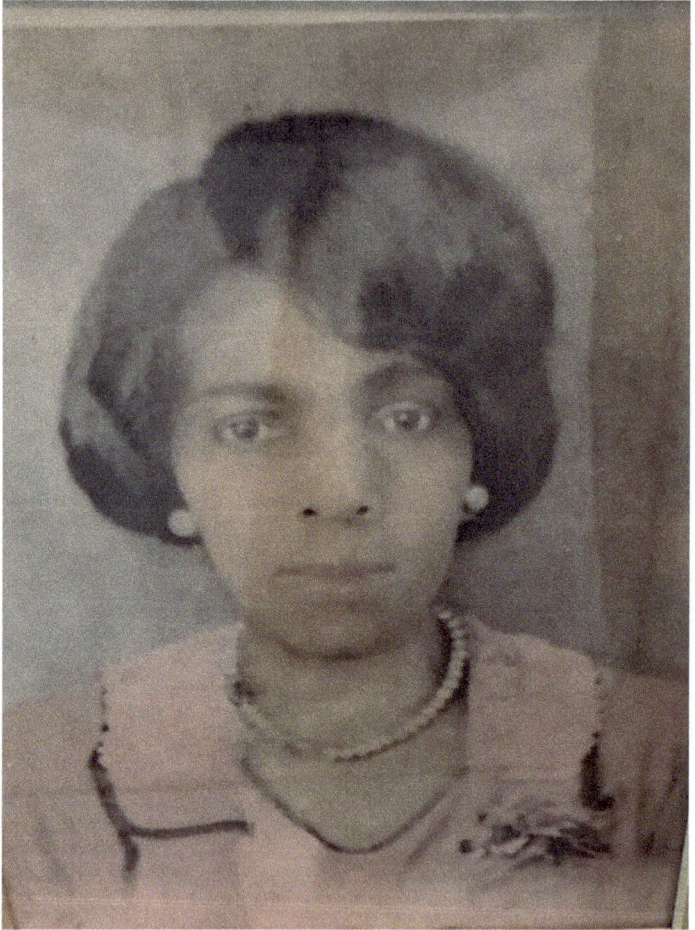

Pictured above is Mama Alberta Mobley Jackson.

Chapter 1: Trip to the Pasture

At five o'clock a.m., before sunrise, Papa went to the pasture to cut firewood that was used for his family and the owners of the farm.

Papa rode on his big, old wagon. He sat in the middle of a long piece of wood, used for a seat. He held a kerosene lantern in one hand. In the other hand, he held the harness straps that were attached to the mule.

Resting on the floor is a rifle, an old-fashioned long saw with a wooden handle on each end, an axe, and a file for sharpening the blades. Also, there are several jars of water covered with lids.

Papa wore an old, faded, blue, long-sleeve shirt and a pair of old, faded, blue loosely fitted overalls. One top pocket of the overalls held a pocket watch. Part of the chain was visible. In the other pocket was a block of chewing tobacco.

Also, he wore a wide-brim straw hat and an old, brown pair of brogan boots.

Two teenage sons, Paul and Ambrose, sat at the end of the wagon. They wore long-sleeve shirts, overalls, brogan boots, and straw hats. Each one held a cloth sack that had a string tied around the top. The sacks contained baked bread and pieces of fried fatback meat.

Chapter 2: Requirements Needed for Rent

1. The first two bales of cotton picked (500-lb. each)

2. Meat: hog meat (hams, shoulder, and fatback) chickens dressed, geese dressed, guineas dressed (Dressed means ready to be cooked.)

3. Lard (fat from hogs)

4. Cornmeal

5. Flour (from wheat)

6. Homemade molasses (made from sugar canes)

7. Wood prepared for the fireplace

8. Vegetables

9. Cantaloupes

10. Watermelons

Chapter 3: Farming Equipment Used

1. A mule-drawn hay rake was a wide two-wheeled implement with curved steel or iron teeth usually operated from the seat mounted. It was used to harvest hay.

2. A mule-drawn harrows was a machine with spike like teeth or upright disks, drawn chiefly over plowed land to level it.

3. A mule-drawn planter was used to plant cotton seeds, corn seeds, wheat seeds, and sugar cane seeds.

4. A mule-drawn plow was a device which allowed the farmer to firmly hold two wooden handles of the plow. Also, in his hands were harness straps attached to the mule, so it could move in the right direction. As the farmer held the handles, he walked behind the mule-drawn plow keeping the

blades to the right soil depth to create a furrow.

5. A shovel was used to dig holes and scoop up dirt. Also, the manure of the mules and cows was scooped up and used for fertilizer.

6. A long handle saw was used to cut down trees in the pasture.

7. A short handle saw was used to cut thin branches.

8. An axe was used to split the word, so it could be used for the wood stove and fireplace.

9. The pitchfork was used to scoop up the hay.

10. A hoe was used for weeding and thinning out the small cotton plants, thinning out the small corn plants, and working in the garden.

Chapter 4: The House

The unpainted wooden house was located about two miles from the main highway. It had two outside doors. One was at the front door, and one was on the back porch. There were no locks nor knobs on the doors. Instead, there were two long thick pieces of wood about four inches wide and four feet long. The boards were placed between two big hooks on each side of the door. When the family had to leave the house, someone had to put the boards in place and then crawl out the window. To get back into the house, someone had to go through the window again and remove the boards. A few years later when my brothers were working in town, they put locks on the doors.

The five windows had neither screens nor locks on them. They were usually raised during the summertime and closed during the wintertime.

The front porch was almost the length of the front of the house. The porch roof was supported by four medium-sized posts. Two six-feet wide backless benches were on the porch. There were three wooden steps. Years later, someone gave us a bench with a swing on it. It stayed at the end of the porch for a long time.

On the back of the house was a semi-enclosed porch. The back of the porch was a wall that was connected to the kitchen.

The right side of the porch was a wall. It was connected to the back of the house. The back door entered the kitchen.

The porch was covered with part of the roof. It was supported by two wooden posts. The low porch had no steps. A long shelf was built between two posts. On the shelf was: a bucket of water with a dipper in it, a tin wash pan, and cakes of lye soap that Mama made.

The tin tub on the floor had a rubbing board in it. Mama used them to wash the clothes. The big barrel stored the dirty clothes. The big trunk stored the clean clothes.

Two long narrow pieces of wood were connected to the wall next to the kitchen. On the boards were many protruding nails. Burlap bags hung on the nails. In the bags were: dried peaches, dried apples, dried beans and peas, onions, white potatoes, and a variety of planting seeds used for the garden.

Chapter 5: The House Interior

The five windows were covered with curtains that were handmade with fabrics from flour sacks or quilt pieces stitched together. They were attached to the window frames with nails.

All the rooms and floors were covered with unpainted horizontal wood planks. When one walked through the front door, the front room was the first room in which the family entered This room was called the front room. The fireplace had a cement outer hearth. The room was crowded with an old-fashioned upright piano, an old sofa, and an armchair.

A bedroom was on the left side of the front room. It held three standard beds. The metal heads of the bed frames were old and unpainted. Each bed had a straw-filled mattress. Mama refilled them once per year after the wheat fields had been harvested. The fabrics from flour sacks were used to make sheets and pillowcases. The pillows were stuffed with cotton and old clothes.

Another door in the front room led to another room. This room had one standard bed in it. The metal head of the bed frame was unpainted. The room had a large fireplace. Sometimes Mama cooked on it. Also, during the winter months, the family members ate their meals in front of the fireplace. The younger siblings sat on quilts on the floor. The older siblings sat on chairs.

At bedtime, Papa extinguished the fire in the fireplace to prevent the house from catching on fire. Mama used a heavy cast iron pot and big skillet to cook on the fireplace too.

One door in this room led to the last bedroom. This bedroom had three standard beds. The bed frames had no heads on them.

The small, crowded kitchen was the last room. It had a wood stove, an ice box, a table and chairs, and a white cabinet that stored the food that had been canned in jars. When there wasn't enough space to store the canned food, Mama stored them under her bed. Some of the canned food included blackberries, peaches, pear preserves, and cooked pork sausages in jars. Also, the canned food did not spoil.

One large bag of flour and meal, a bag of yellow grits, several gallons of homemade molasses in jars, and a churn of buttermilk were seated on the floor. Also, a bucket of water with a dipper in it was on the table.

Chapter 6: Front and Back Yards

The front yard had two big oak trees. They were about one hundred feet apart. Mama planted the saplings when the children were very young. A row of fenced hedges was behind the trees. The space between the trees and hedges was a flower garden that Mama enjoyed growing.

Pots of flowers sat on the porch and on the ground in front of the porch. The names of some of the flowers were: sunflowers, zinnias, petunias, marigolds, elephant ears, four o'clocks, portulacas, and ferns.

About nine or ten feet away from the right side of the house were long single clotheslines. They ran from the front yard to the backyard. Several nails securely held the wires in place on four tree trunks (used as posts).

After the washed clothes were placed across the wires to dry two or three heavy sticks held up the clotheslines. There were no clothes pins.

In the backyard, near the end of the clothesline was a large potbelly cast-iron cauldron. It was used to boil water. When the hogs were killed, the fat of the hogs was boiled in the cauldron to get lard. Mama also used it to make lye soap.

About twenty feet behind the back porch was the ground water well. Circling the well was a raised wooden square floor. A square wooden box, about five feet high and open at the top, sat atop the floor. The hole in the ground was seen from the center of the box. A removable wooden lid covered the square box. On each side of the square box were two strong vertical posts that were connected to the floor. Another strong post was connected horizontally to the vertical posts. The strong wooden bucket sat atop the square wooden box. A tough rope, securely tied in a knot on both sides of the bucket's handle, was connected to the long rope that was wrapped around the center of the horizontal top. Only the adults were permitted to draw water from the well.

Chapter 7: The Barnyard

The front of the barnyard was about twenty-five feet from the back of the groundwater well. A barbed wire fence completely surrounded the big barn that sat in the right back corner. The big horizontal wood plank gate was the only entrance and exit for two mules, three cows, chickens, geese, and guineas.

Two mules and three cows lived in the barn. The hens laid their eggs and hatched their chicks in the barn. Hay, corn, and corn husks were stored in the loft. The cows and mules were fed hay and corn husks. The cows grazed in the pastures, too. The chickens, turkeys, geese, and guineas ate corn.

The hog pen was in the back of the barn, but it was outside the fence. Papa usually raised two or three hogs each year. When all the cotton had been picked, the hogs were killed for meat during the months of January.

Approximately twenty feet to the left of the fence was another barn. The front section was used for the smoke room. This is where the hog meat was cured with salt and stored. In back of the smoke room was where the cotton that had been picked was stored until it was taken to the gin house.

The outhouse or outside toilet (sometimes called) was behind the barn that stored the meat and cotton. It was about six-feet tall and four-feet wide. Also, it had a roof on it.

On the inside of the house, a night pot (sometimes called) was used in place of the outhouse.

An area near the second barn resembled a four-foot teepee. Papa called it the potato hack. It was used to preserve the sweet potatoes during the winter months. After the sweet potatoes were harvested and dried in the sun, Papa built a potato hack. First, he spread a thick layer of hay in a circle on the ground. Second, he placed the sweet potatoes on the hay. Third, he spread another thick layer of hay over the potatoes. After this was done, he placed a four or five-foot post in the middle of the potato bed. Then he placed the cornstalks in an upward position around the bed. They were guided to lean towards the central post. Next, he covered the cornstalks with hay. Finally, he carpeted the hay with dirt (soil). A three or four-inch opening around the post allowed someone to easily reach the potatoes.

The wood pile was located on the same side of the potato hack. Its name was exactly what it was. On the wood pile was a four-foot high block of wood. It was called the chopping block. Papa used it to chop the wood with an axe. The wood was needed for the wood stove and the fireplace. After the wood was cut, it was stacked on the back porch. The chips of the wood were gathered to help start the fires.

Also, on the inside of the barnyard was a six-foot long, three-foot wide trough. It was close to the edge of the fence and almost in the middle on the inside of the fence. This was where the mules, cows, chickens, turkeys, geese, ducks, and guineas got their supply of water.

On the right side of the trough was a huge block of salt on the ground. The mules licked the salt to prevent them from getting diseases. The cows licked the salt to prevent diseases and to help them produce milk.

Chapter 8: Birthday Celebrations

The first time I was reminded about my age, I was ten years old. One night during the Revival Meeting at our church, I sat on the Mourner's Bench in front of the pulpit. At that time, I testified that "I believed in Jesus Christ, and I wanted to be baptized."

One day later when we were home, I became angry with one of my siblings, and I called her a fool. When Mama heard about it, she reminded me that the Bible says I must never call anyone a fool. She told me I would get baptized when I was eleven years old. I cried because I wanted to be baptized like the other children. Today I realize that Mama was right. When I raised my children, I never forgot what she told me.

During the time we lived on the sharecropping farm, there were no birthday celebrations for anyone. In fact, a birthday celebration was not a part of our culture.

Now that I am an adult, I think the most important thing that happened to us was that our parents taught us how to pray the Lord's Prayer and how to live a Christian life.

During those days, our church services were conducted twice a month. However, when the time came, Mama and Papa made sure we were there. When we walked to church, the trip to and from church was eight miles. When my sister, Georgia Ree, bought a car, we rode to church.

Chapter 9: School Time

Cotton planting time started in April. After the cotton plants got a certain size, they had to be weeded and thinned out. That was done row-by-row with a hoe. In fact, we stayed in the fields from 6:00 a.m. to 6:00 p.m.

When we finished our fields, we went to work for a white man to hoe his cotton. Mama took five of us with her. He paid each of us $2.00 per day. At the end of five days, he paid Mama $60.00 for the week. When we picked cotton for the same man, we were paid $2.00 per 100 pounds.

Cotton picking time started in August. It continued until December. After the cotton-picking time ended, we were allowed to attend school. We attended school for three or four months of the school term.

When we did attend school, we had to walk for two miles there and two miles back. When it rained, we stayed home.

After most of the siblings had left the farm, Papa needed the three younger girls to help him with the seed planting. When Mama saw our tears, she said, "George, let them go to school. I'll help you." Mama wanted her children to get a good education. She constantly told us, "Sit down and get your lesson."

Papa knew what he had to do so his family could have a place to live. If he had a lazy bone in himself, we didn't see it.

Chapter 10: Christmas Time

For many years on Christmas morning, my siblings and I looked beneath and undecorated cedar tree and saw several brown paper bags. When we saw them, we knew what to expect. Each bag contained one apple, one orange, two Brazil nuts, two walnuts, one small box of raisins, and one stick of mint candy. The apples and oranges were big and juicy. We were glad to get them because we only got them once a year. In fact, I don't recall getting candy until Christmas time.

When my sister Georgia Ree started teaching school, she gave us something different, but the brown paper bags were always there. During that time, black teachers weren't paid as much as white teachers, but she did what she could do to help her family.

When I think about those times today, I realize that my parents loved us. They did all they knew how to do. In fact, I am proud to have had them as my parents.

Illustration above is our Christmas tree and brown paper gift bags

Chapter 11: Picking Blackberries

Every year during the month of June, Mama took some of the teens with her to pick blackberries. We had to walk about 1½ miles to get to the thorny dense shrubs.

The only problem with picking blackberries was the chiggers that were on them. Chiggers are a tiny larva of mites. When they lodged on the skin, they caused irritating itching.

Because of this, Mama made sure we were dressed properly. We wore long sleeve shirts, long pants, socks and shoes, and a straw hat. Also, each of us took a stick with us in case there was a snake nearby.

When we arrived home, our tin buckets were full. We changed our clothing, washed our hands and faces, and ate breakfast.

The breakfast consisted of biscuits sopped in homemade molasses, fried fatback meat, and water or buttermilk to drink.

When we finished with breakfast, we had to sort out the bad blackberries from the good ones. Then we washed the good blackberries. After this was done, Mama put the blackberries in a big pot and boiled them for a few minutes.

While the blackberries were hot, Mama used a dipper and put them in sterile gallon jars. Then she put a sterile lid on each of them.

For supper that night, we ate a bowl full of blackberry cobbler. Also, during the winter months, we enjoyed delicious blackberry cobblers.

Chapter 12: Milking the Cows

When I was about eleven or twelve years old, it was my turn to manually milk a cow. Since there were three cows to milk, two of my siblings shared the job with me.

The cows were very attentive when we milked them. I think they were glad to be relieved of the milk that filled their teats. The job was not difficult, but we didn't like for the cows to swing their wet urine tails on us while we were milking them.

When we were going to school, we milked the cows early. After we did that, we drove the cows to the pasture to graze during the day. After we were dressed for school, we walked about a half a mile to the highway and caught the bus. Black students started riding the buses in 1955. Before then, we had to walk to school.

In the evening, we went to the pasture to bring the cows back to the farm. They were at the gate waiting for us. After they were in the barnyard, they drank the water that was prepared for them. Water was drawn from the well. In the barn, they ate cornhusks. The mules ate the hay, and the hogs and chickens ate corn.

Chapter 13: Processing the Cow's Milk

After the cows were milked, Mama used a sieve to strain the milk. Then she poured the milk into a 17-inch churn. She put the lid on the churn and let the milk set for a day or two. When the milk turned to curds, Mama used a dasher and beat the milk until butter surfaced on the top. Then Mama used a slotted spoon and removed the butter. The buttermilk was used for drinking. Also, it was used to make delicious biscuits and cornbread. The butter was used to make cakes, pies, and other desserts. The milk that was not churned was used for cakes, pastry, and pies. Egg custard was a delight each Sunday morning.

Chapter 14: Flies

The barnyard was close to the back of the house. It housed chickens, geese, cows, and mules. The hog pen was in the back of the barn. Therefore, one can conclude that flies were inevitable, and they were.

During the summer months, Papa rose early, started a fire in the wood stove, and used his spray gun to kill the flies. When Mama awakened, she swept up the flies and started breakfast.

Flies were always unwanted visitors. However, we were able to adjust day-by-day, week-by-week, month-by-month, and year-by-year.

Chapter 15: Clothes

When I was a sixth grader, my sister, Georgia Ree, who was a teacher in the all-Black elementary school, gave my sister and me a Christmas present. They were corduroy jumpers with blouses to match. My sibling's jumper was burgundy. The blouse was white with burgundy stripes. My jumper was royal blue. The blouse was white with royal blue stripes. That was the first time I received new clothes. Of course, there were the traditional brown paper bags under the finally decorative cedar Christmas tree.

Whenever my elder siblings outgrew their clothes and shoes, there were younger siblings ready to receive them.

Mama washed and iron clothes for two white families. She was paid $2.00 for each job. In addition to this, they gave Mama old clothes for our family. She washed and ironed those clothes. Mama was a master for handling scissors, thimbles, needles, and thread, also. If there were alterations to be made, they were done. We were clean when we went to church and school.

After the elder siblings left the farm, they had better jobs. Not only did they buy clothes for the younger children, but they did not forget Mama and Papa.

Chapter 16: Quilting Time

After the cotton-picking season was over, Mama busied herself making homemade quilts for the beds. A seamstress that she knew gave her straps of cloth. Mama handstitched the multicolored straps in squares. (She did not have a sewing machine until years later.) When she had enough squares to make a quilt, she sewed them together. Then it was large enough to cover the top of a bed.

The next thing that Mama did was handstitched the washed ivory-colored cloths that was used to sack flour. She made the sheet the same size as the quilted pieces.

When the ivory-colored sheet was ready, Mama placed it on the clean floor. Then she covered the entire sheet with unseeded cotton that Papa got from the cotton gin.

After that was done, Mama put the quilted pieces covering over the cotton-covered sheet.

While the sheets were on the floor, Mama placed four long pieces of wood at the end of each side of the incomplete quilt. Next, she used a hammer and nailed each end of the boards together.

After that was done, she nailed the incomplete quilt to the boards. This allowed the incomplete quilt to be stretched out from side-to-side.

Hanging from the ceiling were four long ropes. They were used to hold each end of the boards in place. The ropes were used to raise the incomplete quilt to a certain height. Also, the two sheets stayed in place while Mama and her neighbors handstitched it. Sometimes, they stood up, and other times they sat on chairs. I heard them talking and laughing. When Mama was quilting alone, I heard her singing some of the songs we sang at church. I really think that quilting was a peaceful time for her.

When the hand stitching was complete, the quilt was detached from the boards. The final step was to hem the border around the entire quilt.

On several occasions, I saw Mama give visiting relatives a quilt. I could tell that they were amazed as well as thankful.

Quilting continued until springtime. Then gardening and planting different crops began.

Illustration above is Mama quilting

Chapter 17: Mama's New Washing Machine

Believe it or not, the washing machine is still sitting on the enclosed back porch of the family's house. Each time I see it, I think about Mama.

In 1957, my sibling, Christine, the eldest, bought Mama a new washing machine. I can visualize how Mama carefully used the washing machine with happiness and a relief that she would not be bending her back and scrubbing on the washboard.

While Mama was packing the dirty clothes in the washing machine, my siblings and I drew water from the well and carried it to the back porch where the washing machine was positioned.

After the appliance was filled with water, we filled the two big tin tubs with water, so the clothes could be rinsed.

When Mama decided that the clothes were clean, she pulled the lever and stopped the appliance.

At the top of the washing machine were two round wooden rolls touching each other. When Mama pulled the lever next to them, the rolls started turning sort of fast. When Mama carefully placed the clothes so they could touch the rolls, the clothes went through the rolls, and water was squeezed from them.

After the clothes went through the ringer, my siblings and I rinsed the clothes in the big tin tubs and put them on the clothesline to dry.

During the cotton-picking season, my family usually picked cotton on Saturdays from 6:00 a.m. until 12:00 noon. If clothes had to be washed, Mama did it then.

Pictured above is Mama's new washing machine.

Chapter 18: Recreation on the Sharecropping Farm

Working on a cotton farm required hard work. Every job had to be done manually. There were no modern machineries that are used nowadays.

First, the cotton seeds had to be planted. Next, the cotton plants had to be thinned out and weeded. Insecticide had to be used to kill the boll weevils.

Cotton picking began in August and ended around the early part of December. Most of the recreational time was spent on Sundays. Papa made a swing in the oak tree. Other activities included: jumping rope, shooting marbles, a hopscotch game, a checker game, and a Tic-Tac-Toe game. Also, we had a radio that played one station.

One day our brother brought home a television. If my recollection is correct, we could only see two channels, but we enjoyed them.

Chapter 19: Bicycles and Skates

When the eldest brother, Christopher Columbus, got a job laying bricks with a friend of the family, he bought a new bicycle that he needed to take him to and from work. He kept that job until he was drafted for the army.

When Columbus went to the army, the three brothers rode the bicycle until it was no use to them. Frances Virginia and Betty Jean were given a blue bicycle for them to share when they were teenagers.

When Gloria, the youngest, was ten years old, and I was thirteen, we were given a red bicycle to share.

Mama didn't let the girls ride the bicycles on the highway. We rode them in the yard and on the dirt road.

When Gloria, the youngest, was given a pair of skates, she took them to school. During recess time, she learned how to skate on the paved walkway.

Chapter 20: A Supplementary Income

During the years my family lived on a sharecropping farm, the owners of the farm received most of the money made for the cotton crop. The farmer got what was left. Usually, the farmer's debt was incomplete at the end of the cotton season. In fact, the debt increased year after year. Therefore, the farmer's debt continued.

Papa and Mama raised large gardens for more than one purpose. When the vegetables were plentiful, Papa packed his wagon every Saturday and sold them to the people in the city. In addition to this, he sold cantaloupes, watermelons, peanuts, sweet potatoes, young calves, and piglets.

The money Papa made was used to buy things that he could not grow on the farm. For example, Papa drank coffee and chewed tobacco. The mules needed to be shoed once per year. Worn out bridle straps had to be replaced. I saw Papa axle oil the wagon wheels. Sometimes a wheel had to be replaced. Walking in the fields daily throughout the month will require a new pair of shoes (usually once per year).

Mama needed things to help her prepare the meals. Some of them were: black pepper, table salt, sugar, flavors, and spices. She also needed laundry detergent.

Since there was an ice box in the house, Papa bought a block of ice from the store every Saturday. Mama made iced tea on Sundays, for Thanksgiving Day, and for Christmas dinner.

Chapter 21: About Meals

During the time we worked in the cottonfields, we began early in the morning. After we worked for a few hours, we went to the farm house to rest and eat lunch. The lunches included one of the following: cornbread and buttermilk; biscuits sopped in molasses; or stickers which were similar to cinnamon rolls, but Mama baked them in a big pan. We drank plenty of water or buttermilk. We ate a lot of molasses bread too. I can't recall how Mama made it, but it was delicious.

When we returned to the cotton fields, the next break we had was a brief pause to drink water. My youngest sibling, Gloria, carried the bucket of water (with a dipper in it) to each person. Gloria was the only sibling who did not work in the fields. She was frightened of worms. Even when she was carrying the bucket of water, she was cautiously looking for a worm.

For Sunday breakfast, Mama made several delicious egg custards with homemade pie shells. According to my recollection, her main ingredients were: eggs, milk, butter, and sugar. They were flavored with dried orange peels, vanilla flavor, or nutmeg spice. When I became an adult, I attempted to make an egg custard, but I was unsuccessful.

For Sunday dinner, we ate fried chicken or chicken with dumplings, white potatoes, a green vegetable (if they were in season). For dessert, it was a cobbler, or a sweet potato pie. The beverage was iced tea.

For Thanksgiving dinner, Mama cooked pumpkin, baked goose or hen, collards or turnips or turnip greens, cornbread, plain pound cake for dessert, and iced tea. Sometimes we didn't have ice with it. We learned about Kool-Aid years later.

For Christmas dinner, Mama baked a goose or goose dumplings, vegetables, yams, and dried apple pies (another of her specialties) cornbread and iced tea.

When I think about the meals we ate during the wintertime, my vivid memory focuses on one of the following: 1) cornbread in a cup of buttermilk, 2) cornbread sopped in homemade molasses and seasoned with oil from fried fatback meat, a slice of fried fatback meat, and buttermilk to drink, 3) biscuits sopped in homemade molasses seasoned with oil from fried fatback meat; a slice of fatback meat, and a cup of buttermilk, or water to drink.

Chapter 22: Social Security

On August 14, 1935, President Roosevelt signed the Social Security Act in law.

In 1950, fifteen years later, Social Security was approved for domestic workers and sharecroppers.

Papa and Mama started receiving their Social Security benefits in 1960 (twenty-five years later). Their combined monthly pension was about $100.00.

Chapter 23: The New House

During the time when Paul and Ambrose worked away from the farm, they realized that one day soon, Mama and Papa would reach the age when they could no longer be able to do farm work. When that stage in their lives occurred, they would have to leave the farm. As a result of this, they would be homeless.

In 1954, Paul and Ambrose paid a down payment on two acres of land. It took them one year to finish paying for the property.

In 1955, Paul and Ambrose laid the foundation and started building the block walls. As the work progressed, the block walls were covered with bricks. They only worked on weekends since they attended to their jobs during the weekdays.

A short time later, Ambrose was drafted in the Army. Paul and Columbus continued to work on the house during the weekends.

When Ambrose was discharged from the Army after he had served for a little over two years, he continued to help Paul and Columbus work on the house.

Elizabeth, the second oldest sibling, was working in Newark, New Jersey. She continued to send money for purchasing building supplies.

In 1961, six years later, the house was completed. Mama and Papa plus five siblings moved in the house. Unfortunately, Mama only lived in the new house for two years. A massive heart attack ended her life. Papa continued to live in the house for twenty-seven years.

Pictured above is the new house built by Paul, Ambrose, and Columbus

Chapter 24: Grandma Lizzie's Secret

Our family was blessed to enjoy Grandma Lizzie, our maternal grandmother. Her husband died before any of us were born. Papa's parents were deceased also.

Grandma Lizzie's small frame was about 5' 5" tall. She always stayed the same size. If I were to guess, she looked as if she weighed about 130 pounds.

Grandma Lizzie grew vegetables in her garden. She never put a piece of hog meat in her mouth. She said, "Hog meat would make my blood high." She raised chickens for meat and eggs. She grew strawberries too. Mama gave Grandma Lizzie: milk, butter, and vegetables that she didn't have in her garden already.

When two or three of the younger siblings in our family spent the night at Grandma Lizzie's house, we knew that our breakfast beverage was going to be warm sweet mint tea. She grew mint plants in her garden.

Immediately after Grandma Lizzie finished her meal, she put a powdery substance in her lower lip. Then she kept a tin cup nearby, so she could spit in it.

When I was older, I learned that she dipped snuff. She continued that habit until her health failed.

When my family was able to get a car, we took Grandma Lizzie to church with us. Before we got a vehicle, we walked to church. It was an eight-mile round trip.

Every second and fourth Sunday, after the church services, Grandma Lizzie ate dinner at our house.

When Mama asked her to stay overnight, she told Mama she wanted to sleep in her own bed. In fact, Grandma Lizzie never slept overnight in our farmhouse.

When we moved in the new house, Grandma stayed overnight several times.

After I graduated from high school, I coughed up the nerve to ask Grandma Lizzie why she never spent the night at our farmhouse.

Suddenly, she laughed and said, "Baby (a name she gave the children sometimes), That house was full of ghosts. I could see them come in through the doors and go out through the doors. They would walk from room to room. They never bothered nobody."

I asked, "Grandma Lizzie, why didn't you tell somebody?"

Grandma Lizzie asked, "Where would your mama and daddy go with thirteen children?" She then said, "They had to raise y'all." She added, "If you children would've knowed about it, you would've been scared."

Grandma Lizzie lived thirty-six years after Mama passed. She was ninety-nine years old.

Pictured above is Grandma Elizabeth Mobley, affectionately known as Grandma Lizzie.

Chapter 25: Siblings

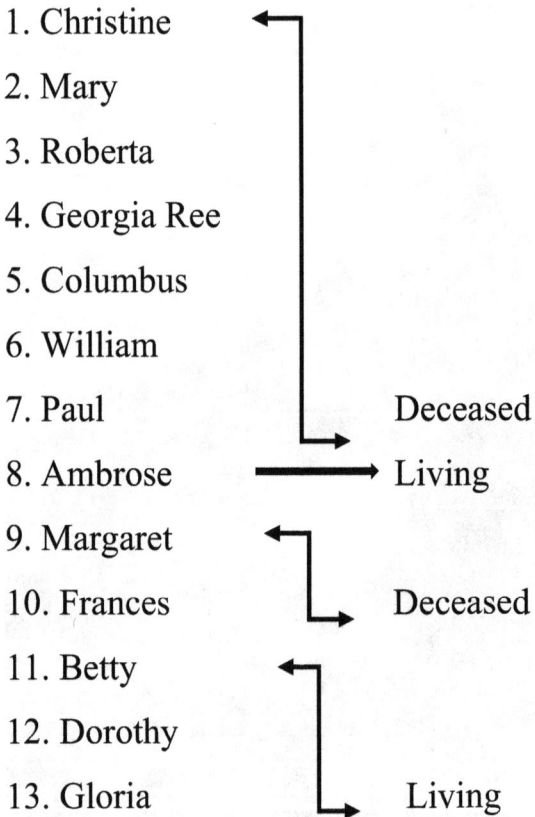

1. Christine
2. Mary
3. Roberta
4. Georgia Ree
5. Columbus
6. William
7. Paul   Deceased
8. Ambrose   Living
9. Margaret
10. Frances   Deceased
11. Betty
12. Dorothy
13. Gloria   Living

The owners of the sharecropping farm gave Mama and Papa the following names:

1. Cleo Christine
2. Mary Elizabeth
3. Georgia Ree
4. Roberta Beatrice
5. Christopher Columbus
6. William Theodore
7. Paul Hoover
8. Ambrose Hampton
9. Margaret Helen
10. Frances Virginia
11. Betty Jean Rebecca
12. Dorothy Lamour
13. A member of the family named the youngest.

His Best Crop Was Children—13 Of 'Em

Dennis, Bob. "His Best Crop Was Children – 13 Of 'Em". *The Charlotte Observer*. 02 October 1967, p. 1B

Chapter 26: My Family

Once upon a time, there lived Papa, the father, and Mama, the mother. They lived on a sharecropping farm with their thirteen children. The house they lived in had four small rooms and one smaller room for the kitchen. For several years, they had no electricity.

Christine, the oldest sibling graduated from high school. After that, she obtained a job in a factory. She used her money to help take care of her siblings. Also, she bought the first car in the family. Later, she left that job and enrolled in a two-year college to study to become a teacher. After she taught school for two years, she decided she wanted to enroll in the practical nursing program. She continued with the nursing career until she retired. She spent the rest of her life being a born-again evangelist.

Mary Elizabeth, the second sibling, continued in school until she graduated. After graduation, she did domestic work for ten years. Then she decided to go to a four-year college to study to become an elementary school teacher. She taught in the public school until she retired.

Roberta, the third sibling, graduated from high school. Then she got married. She and her husband raised four children.

Georgia Ree, the fourth sibling, also graduated from high school. She was the second one to graduate from college. After she received a teaching degree in elementary education, she taught in the public schools until she retired. She bought the second car in the family.

Columbus, the first male sibling and the fifth child, dropped out of high school and started working as a bricklayer. Also, he spent a short time in the military. He and his wife raised six children. Before he retired, he had the reputation of being a master bricklayer.

William, the second male sibling and the sixth child, was the first male in the family to graduate from high school. After graduation, he began working with his brother Columbus, laying bricks. Later, he was drafted to serve in the military. After he was discharged from the Army, he got married. He and his wife raised five children. William was the first minister in the family, and he was the first sibling to get his driver's license.

Paul was the third male and the seventh sibling. He was the second male in the family to graduate from high school. After graduation, he worked with his brother, Columbus, laying bricks. He was the first male in the family to attend college. After a short time there, he decided to continue to work as a bricklayer. Also, he started to learn the carpentry trade. He was known as a master carpenter. Paul was the second male in the family to become a minister. He and his wife raised three children.

Ambrose, the fourth male child and the eighth sibling, dropped out of high school in the ninth grade. He left the farm and began working for a lumber company. After he left that job, he started laying bricks with his three brothers: Columbus, William, and Paul. When he reached the age of twenty-one, he was drafted in the Army. He stayed in the military a little over two years. After he was discharged from the Army, he received his GED (high school graduation equivalency diploma). Ambrose was the third male in the family to become a minister. He and his wife raised three children.

Margaret was the ninth sibling. After she graduated from high school, she enrolled in a Beauty Culture School and studied to become a beautician. When Margaret completed the course, she relocated to New Jersey to live with her sister. Mary Elizabeth. She worked as a beautician until she became ill and passed. She was survived by her husband and two young children.

Frances was the tenth sibling. When she graduated from high school, she enrolled in a four-year college and majored in

Biology. After she taught school for approximately five or six years, she was killed suddenly in a single auto accident. She and her husband had one daughter.

Betty Jean, the eleventh sibling, was a member of the varsity basketball team during her high school years. After graduation, she enrolled in a four-year college. While she was there, she studied to become an English teacher. She taught in the public school until she retired. She and her husband raised three children. Also, she had four stepchildren.

Dorothy, the twelfth sibling, was the first in the family to learn how to play the piano well. While she was in college, she studied to become an English teacher. She taught in the public school for thirty years. Also, she was the musician for the youth choir of her church for approximately thirty-five years. In addition to this, she is the first sibling who is the founder of a Christian School. It has been operating for twenty-four years. She and her spouse raised five children.

Gloria, the thirteenth sibling, graduated from high school and enrolled in a four-year college. There, she studied to become an elementary education teacher. After graduation, she relocated to Newark, New Jersey to live with her sister, Elizabeth. She taught in the public school for several years. After she left the teaching profession, she obtained her license to be a CNA (certified nursing assistant). She continued with this position until she retired.

Of the 24 great-nieces and great-nephews, all obtained college degrees. One has a career in the military. Three have careers in education. One has a law degree, and one presently is in medical school.

Photo: Taken on January 23, 1964 on the day of Mother Alberta Mobley Jackson's funeral

Top Row (Left to Right): Mary Elizabeth Jackson, Frances Virginia Turner, Margaret Helen Jackson

McCullough, Christopher Columbus Jackson, Cleo Christine Jackson Blake, Georgia Ree Jackson, Roberta Beatrice Jackson Campbell, William Theodore Jackson (Rev.)

Second Row (Left to Right): Dorothy LaMore Jackson McQueen, Rev. Paul Hoover Jackson, George Washington Jackson (Father), Betty Jean Jackson Evans, Rev. Ambrose Hampton Jackson, Gloria Berenes Jackson Miles

Husbands: Frances (John), Margaret (Jonathan), Cleo (James), Georgia (William), Roberta B. (Lindsey), Dorothy (John), Betty (Spofford), Gloria (Vincent)

Wives: Columbus (Helen), William (Pearlena), Paul (Janie), Ambrose (Phyllis)

About the Author

Dorothy Jackson McQueen is a native of Chester, South Carolina. She spent 30 years teaching in the public schools of Marlboro County, South Carolina. Also, she was named Teacher of The Year of Marlboro County High School for 1990-91.

Mrs. McQueen is the Founder of DEPICAM Christian School in Bennettsville, SC. It is still in operation today after 27 years.

Mrs. McQueen is a member of the Macedonia Missionary Baptist Church. She has served in many capacities: Youth Director, Youth Choir Musician, Vacation Bible School Director, Chairman of The Deaconess Board, and Sunday School Teacher.

Mrs. McQueen and her late husband, Deacon John Troy McQueen, are the parents of four children and three grandchildren.

Author
Dorothy J. McQueen

www.ingramcontent.com/pod-product-compliance
Lightning Source LLC
Chambersburg PA
CBHW052202090426
42741CB00010B/2371